THE SEA WITHIN

By Peter Kreeft from St. Augustine's Press

Socratic Logic
Socrates' Children: Ancient
Socrates' Children: Medieval
Socrates' Children: Modern
Socrates' Children: Contemporay
Socrates Meets Kierkagaard
Socrates Meets Freud
Socrates Meets Sartre
Socrates Meets Marx
Socrates Meets Kant
Socrates Meets Hume
Socrates Meets Descartes
Socrates Meets Machiavelli
Philosophy 101 by Socrates
Socratic Introdution to Plato's Republic
Summa Philosophica
Platonic Tradition
Philosophy of Jesus
Jesus-Shock
Ocean Full of Angels
If Einstein Had Been a Surfer
I Surf, Therefore I Am

THE SEA WITHIN

Waves and the Meaning of All Things

PETER KREEFT

St. Augustine's Press
South Bend, Indiana

2 3 4 5 6   23 22 21 20 19 18 17

Library of Congress Cataloging in Publication
Kreeft, Peter,
The sea within, waves and the meaning of all things /
Peter Kreeft
p. cm.
Includes index.
ISBN: 978-1-58731-757-6 (hardcover: alk. paper)
1. Meaning (Philosophy) 2. Ocean --
Miscellanea. 3. Ocean -- Meditations.
4. Philosophical theology. I. Title.
B106.M4K74 2006
128 -- dc22   2006002217

St. Augustine's Place
www.staugustine.net

TABLE OF CONTENTS

The Quest   1

Data   3

Magic   5

The Sea as Beloved   9

Childhood   15

Infinity   18

Beaches   20

The History of Sea Love   25

The Sea as Koan   37

Orenda   41

The Joy of Ignorance   44

Water   47

Seeing the Sea with the Third Eye   50

The Sea as Miracle   55

Music	66
Water in Love	73
Aliveness	78
The Dark Side	81
The Sea as Toy	88
The Sea as Heater	93
Time and Power	99
About the Author	101

THE QUEST

Why are we in love with the sea? Everyone knows we are, but no one knows why.

This is a riddle we long to read, but cannot:

Who knows the mighty secret,
The secret of the sea?
I love its beauty passing well,
I love the thunder of its swell,
I love the glory of its play,
The glitter of its feathery spray,
But its secret is hid from me.
Who knows the mighty secret –
What gives the sea its power?

We cannot find the secret,
We cannot find the key.

So writes Susan K. Phillips in "The Secret of the Sea." Hers is a simple question, a familiar question. But suppose we refused to let it go after a few teasing passes. Suppose we pursued this riddle as passionately as a starving hunter pursues an animal when his life depends on catching it. What would happen then?

The question about the sea, like the sea itself, would take us on a long and unhomely journey over fathomless depths, through storms and fogs, and with no guarantee of landfall.

Will you come?

DATA

We always have double data. Our senses give us external data, and our hearts give us internal data. When the two do not line up, we have a puzzle. And our puzzle here is this: Why does the image of God feel such passion for the cold salt water? Why do immortal spirits created only a little lower than the angels fall so desperately in love with a trillion tons of H_2O laced with NaCl?

Most books about the sea are full of external data. They tell you what causes storms, for instance. But they don't tell you what causes our fascination with storms: they don't tell you about the storm within. They tell you how the wind raises waves. But they don't tell you how the waves without raise waves of wonder within.

On this journey (are you still with me?) I will try to find the answer to that question by looking at three sources of data. I will squint through the telescope of my soul at myself and at two kinds of external data, the sea itself and the millions of words written about it, especially by the poets.

But I find my data first in myself, not first in the poets. For if I did not find it in myself, I would not be able to find it in the poets. Just as there are things about a woman that only a woman can understand, there are things about the sea that only the sea within me can understand.

Magic

Our internal data varies from person to person, of course, because personality itself varies. But not totally. Whenever we experience something we think is utterly unique, we usually discover someone else to whom we can say, in amazement, "You too!"

What is the sea to me? If I had to say it in one word, I would use the word "magic." I know it is a cliché, but I mean it literally. The sea transforms my body and my soul like a magic wand. This is the first and most obvious datum to me.

For instance, the sea makes me a Buddha. The word "Buddha" means "an awakened one." No matter how tired I am, the sight and sound and smell and touch and taste of the sea wake me up. I become instantly alert, aware, awake, alive, aroused.

Where does this energy come from? Not from me. It comes *to* me. From where? Obviously, from my magic Mother, who fills the dry sponge of my soul.

I will not pause to justify "Mother" or "She" for the sea. Some things are so certain that they can't be explained. As long as you have to ask, you won't be able to answer. The sea transcends ideologies, whether chauvinist or feminist.

She is magic both to my body and to my soul. I have a seventy-year-old's body, yet she gives me a seven-year-old's soul and a seven-year-old's happiness. After an hour of smiling at her and not realizing I am smiling, my face feels funny. I don't smile much, so when I notice this, my *mind* feels funny. Nothing else has ever done that to me.

She makes me to lie down in green pastures of water by leading me away from the green pastures of land. She leads me beside the still waters of contentment by never being still or content herself. Her restless waves give rest to my soul. She restores my soul. She restores my past by washing it away. She does the impossible. She is a time machine.

Of course, all nature is full of magic: star-stuff becomes a planet, a seed becomes a tree, primordial slime evolves into rational, sophisticated slime. But all that takes time; "sea-and-soul magic" is instantaneous. My soul obeys Sea's magic wand as instantly as the universe obeyed the Creator's word "Be!" (Are these little bangs in my soul echoes of His Big Bang, I wonder?)

As soon as I see waves, I have to stop myself from running madly at them and jumping into them, like a lemming jumping off a cliff. I whack them like a baby elephant on amphetamines. I suck their salty taste and smell like a kitten rooting in her mother's nipples.

When I am by the sea, I notice I am breathing more deeply. I notice that I *notice* more deeply. I notice how the air feels because I notice the sea air. I notice all colors because I notice the color of the sea today.

(Every day she is a little different.) I notice all smells because I notice how She smells today. (That difference is also detectable, but subtler.)

A good lover gives you more energy to love *everything*. A bad lover drains away your energy into himself like a vampire. The sea is a good lover to me.

The Sea as Beloved

Half of humanity shares my love affair
with her. Sea-love is as old as the Viking
poem "The Seafarer":

Often I've suffered in foreign seas,
Night shades darkened with driving snow
   . . .
Yet still, even now, my spirit within me
Drives me seaward to sail the deep,
To ride the long swell of the salt sea-
   wave.
Never a day but my heart's desire
Would launch me forth on the long sea-
   path.

And this same love is as modern as
Masefield:

I must go down to the sea again –
To the lonely sea and the sky
And all I ask is a tall ship
And a star to steer her by

Modern times are not the low tide of sea-love but the high. Our minds may have demythologized the sea, but our hearts have not. We no longer think of it as Mother, or goddess, or spirit, as our ancestors did, but I think we love it even more than they did. Just look at the prices on real-estate ads for oceanfront land. For a century now, beaches have been the most popular vacation spots. (But only for one century. Why?)

How can we pinpoint the power of the sea over our hearts? It's not just swimming that we love. It's easier to swim in a pool or in a lake than in the sea. Look at all the people who come to the sea but *don't* swim in it. Just knowing it's there, just being near it, is their satisfaction. Most visitors to the Alps are not mountain climbers or skiers either. Just seeing her face is enough. It's a Beatific Vision. It's love.

Think about what we see. Gather data. On a summer day at the beach, how many people go in the water? Maybe half. Why do the other half come to the beach? Not for the sun; they can get a suntan inland just as well. If there were no ocean, who would come? It's not just the sand and the sun that they love. They don't go to the desert just to sit on the sand in the sun. It's being near Her that they want. That alone satisfies something in us.

What is that something? It seems to be as deep as the sea itself. "Deep calls unto deep." The sea without calls to our sea within, and the tides of our spirit are pulled by the gravity of the sea. What is this inner sea?

When we're at the place where the sea and the land meet, why do we love to look at the sea rather than at the land? When we're out of sight of land, we can only look at the sea; and when we're out of sight of the sea, we can only look at the land; but when we're at the shore, we have the choice to look either at the land or at the sea. Why do we choose to look at the sea?

The people along the sand
All turn and look one way.
They turn their back on the land.
They look at the sea all day.

So observed Robert Frost. He knew
where the magnetic pole of the human spirit
was located: just offshore, in the waves. But
did he know why? Does anyone?

Kipling made the same choice:

Who hath desired the Sea? Her excellent
    loneliness rather
Than forecourts of kings, and her outer-
    most pits than the streets where men
    gather.

Frost and Kipling see the data, but what is
the explanation?

Maybe the sea stands for Heaven. Alfred
Noyes wrote:

Give me the sunlight and the sea
And who shall take my heaven from me?

But how can the sea symbolize our heavenly home if we are not fish? Why do we feel like Edna St. Vincent Millay, exiled in our own world, the world of people, and more at home in the world of fishes?

> Searching my heart for its true sorrow,
> This is the thing I find to be:
> That I am weary of words and people,
> Sick of the city, wanting the sea.

Yes, we do, but why? Suppose we dared to keep repeating the question? What would we find if we were as brash and bothersome as a five year old, asking not just "Why?" but "Whywhywhywhywhy?" forever until we found it? What is "it"? What is Her power over our hearts?

Carl Sandburg says that "**the sea hugs and will not let go.**" Yes, but what are her arms? Notice the grammar: she is the subject and we are the object, rather than vice versa. How can the sea be active and the human spirit passive under her spell? We act on the sea with our ships, but she acts on our soul with her beauty. We conquer her physically, but she conquers us spiritually. How does she come to have such power?

Is the gravity that draws our soul to the sea in us, or is it in the sea? What is the link between the matter without and the spirit within, between salt water and love? Is it the sea water in our souls, or is it the soul in the sea water? Or is that very question the wrong one, one that could be asked only by a modern mind that has separated reality into mere matter without and mere spirit within, the intellectual albatross of Descartes and the so-called "Enlightenment"? We may have to turn back the cultural clock to solve our riddle.

CHILDHOOD

And this seems to be a clue, for what the sea does to us is to turn back our personal clock:

Backward, turn backward, O time in thy flight!
Make me a child again just for one night.

The sea actually does that. We babble like children or mystics when we contemplate Her:

The sea! The sea! The open Sea!
The blue, the fresh, the ever free!

That's from B.W. Proctor's "The Sea Lover." It's also from your heart and from your memory. Do you remember the very first time you ever heard the sound of the sea, or your very first sight of it? You never, never saw anything so big before! And you never will, until you die. Do you remember the first time you smelled it, and touched it, and tasted its soft, salty wetness? Even if your conscious mind can't remember, your unconscious mind can't forget.

Our love of the sea as adults probably has a lot to do with our memory of our childhood and the sea's power to bring us back to that, or to bring that forward to us, to resurrect the dead, to present the past. Byron knew that:

And I have loved thee, Ocean! and my joy
Of youthful sports was on thy breast to
   be
Borne like thy bubbles, onward: from a
   boy
I wanton'd with thy breakers – they to me
Were a delight! And if the freshening sea
Made them a terror – 'twas a pleasing
   fear,

For I was as it were a child of thee,
And trusted to thy billows far and near,
And laid my hand upon thy mane – as I
   do here.

When I have to leave the sea, I feel like a child parted from her mother and her home. For "home" means "where Mother is." I feel homesick for the sea.

But how can this be? Is not my home here? Am I really the angel of the sea with amnesia? Am I an amphibian? Did the sea cough me up on a beach in a storm? Was I born inside Jonah's whale? Am I a merman? I feel exactly like a merman trying to walk on land. My body has legs, but my soul has fins. Why does my soul feel this way?

I don't know the answer to these questions. I only know that I have to go to the sea because I have to follow my soul, and my soul is there already.

Infinity

Perhaps the sea without, though not literally infinite, is so nearly infinite, so apparently infinite, that it is a natural symbol for the really infinite sea within. Perhaps the almost-infinite sea of water which we love is a mirror of the really-infinite sea of spirit which we are.

The sea is the goal of all the water on earth, every splashing brook and running rivulet and roaring river. It is also the goal of the moving water of our spirits. Our spirit-water follows its love-gravity to its end. We are satisfied only at the sea because we are rivers.

But this still doesn't answer our question. If rivers are natural icons of our restless hearts, and the sea is the natural icon of God, and our hearts are restless until they rest in God, what is it that we long for when we long for God? And why is the sea a natural symbol of that? We know what the sea is made of – water – but what is God made of? "Living water." Life! But what is that?

We know *where* we find what we want: at the sea. But we don't know *what* we want there. We know what we long for – the sea – but we don't know what it is that we long for when we long for the sea.

Perhaps we never will. Perhaps the infinite sea can never fit into finite mental or physical cups. Perhaps all that can be clear here is this: that all that is here can never be clear.

Perhaps this "not-clear" is a clear clue that only God answers our riddle. For "infinity" sounds like a divine attribute. If there were any other solution to our riddle, any solution *here,* wouldn't *somebody* somewhere, some time, have found it here?

Beaches

We love not just the sea but the seaside, the boundary, the beach.

It takes boundaries to make anything interesting. If a picture didn't have the boundary of its frame, it would trail off into vague, boring everything-ness. Life's most dramatic moments are her two frames: entrance and exit, beginning and end, birth and death. Light overcoming darkness is more interesting than endless light. We love boundaries. Robert Frost wrote, **"Something there is that does not love a wall,"** but something else in us *does*.

We love to explore and test boundaries: parental authority when we are children, fashion when we are teenagers, fashionable dogmas when we are scientists or scholars.

So we also love to explore the boundary between land and sea.

Infinity needs a frame to register on our finite minds. Infinity (like the sea) becomes interesting to us only when bounded by finitude (like the beach). If the whole planet were land without sea, or sea without land, it would be as boring as if the whole of humanity were males without females or females without males.

Like male and female, the sea is most interesting at its boundaries because that's where we find the tumbling and turbulence and drama and conflict and lovemaking and ecstasy. The land's point of ecstasy, where it reaches beyond itself into the sea, is the beach. The sea's point of ecstasy, where it reaches beyond itself into the land, is the surf. So the surf on the beach is the most interesting place on earth. Like sex, it is interesting because it is *other*. It divides as it unites, like a hyphen or a breezeway. It is a line, a lightening bolt that shatters our dull, homogenous fog.

This otherness of place (the beach) also

seems to have a power to attract an otherness of time: the very young and the very old love the beach the most. Your first summer at the sea and your last are two of the most poignant times in your life.

Watch the very young at the beach. They play in the Great Mother's untiring waves, and become almost as heedless of time as She is. How remarkable that those who are bored the quickest with everything else – the young, the ones with the shortest attention spans – are always the last to want to leave the beach!

The sea is more than a time machine; she is a time-transcending machine. In fact, she is an eternity-machine. Look at the old; what attracts them to the sea? Surely they are drawn by a sense of eternity there. The sea prepares them for their journey to eternity. They know that the seashore is a natural symbol of death, and they know they will soon be "crossing the bar." They are reconnoitering.

The very young and the very old have one thing in common: not much time separates them from eternity, whether eternity past or eternity future, before birth or after death.

But *why* does the sea naturally symbolize eternity?

It is easier to see why it symbolizes death. Death is life's edge, life's other. That's why death makes life interesting: because it is the other, "the undiscovered country." How much more interesting the world would be if it were flat! Then it would have an edge, and you could fall off it.

Well, the world *is* flat in time, even if it isn't flat in space; and we do fall off the edge – at death.

Even in space there is an image of this; even in space the world really does have an edge. The edge is the sea.

And you *can* fall into it.

And you want to. (Death-wish?)

Besides sex-otherness and space-otherness and time-otherness and death-otherness and land-edge-otherness, the beach also sharpens your consciousness of clothing-otherness. It makes you appreciate both bareness and clothedness. It feels great to peel off all the layers of clothes and run bare and free; and at the end of the day it also feels great to get your clothes on again after you shower. You never appreciate your clothes as much as you do after a day at the beach.

If it is true that the most lovable part of the sea is its land-border, then you don't have to be a sailor to love the sea. In fact, in a way the oceanfront landlubber appreciates the sea *more* than the sailor does. James Russell Lowell knew this when he wrote that **"The sea was meant to be looked at from the shore, as mountains from the plain."** And Thoreau knew it when he wrote in "The Fisher's Boy,"

I have but few companions on the shore.
They scorn the strand who sail upon the
**   sea.**
Yet oft I think the ocean they've sailed
**   o'er**
Is deeper known upon the strand to me.

The History of Sea Love

Let's explore another kind of data: our history, the history of our thoughts and feelings about the sea. Perhaps our past can shed some light on our present.

Mother. That's what the sea was for all of us until a very few millennia ago.

To our ancestors who made the myths, she was She: perhaps a goddess, perhaps a demon, but certainly a spirit, and "always a woman to me."

To the Jews, Christians and Muslims who came after the pagan mythmakers, she was no longer a goddess but she was not yet a mere chemical. She was an icon. She was de-deified but not yet demythologized. She was not the divine Artist, but she was not raw paint and canvas either. She was the art. She was no longer an object of worship, but she was still an object of wonder.

She was alive. Her waves clapped their hands with joy when they felt their Creator's breath blowing over them. Her performances were all command performances. She obeyed her Commander's laws by keeping her bounds, and our ancestors were wise to wonder at this: that such a huge and lively thing should stay so meekly in its world-wide cage behind invisible bars. They found her peace more surprising than her storms, because they were closer to seeing her as a spirit than as a chemical.

When the Lord answered Job out of the storm (Job 38:1, 8–11), He said:

Who shut in the sea with doors
When it burst forth from the womb,
When I made clouds its garment
And thick darkness its swaddling band,
And prescribed doors for it,
And set bars and doors,
And said, "Thus far shall you come and
  no farther,

And here shall your proud waves be stayed"?

We religious crazies, we Jews and Christians and Muslims, know a secret about the sea that our secular friends do not know. They have visions of a day, millions of years away, when the last man and the last woman will look on the sea for the last time. In their scenario, the sea will outlast us. Just as its life extended far into the past before the coming of this upstart species *homo sapiens,* so it will extend far into the future after that species' final exit from the earth. That is their scenario. But we know a secret about the future that they don't know: we know that we will outlast the sea. In fact, we will outlast the stars. When all the stars are dead, each one of us will be alive.

Knowing that frees us from the ancient temptation to worship the sea, or the earth, or the stars – and the modern temptation to worship even more ephemeral things like society or politics or ideology.

We God-people stand in the middle era,

between the ancients, who worshipped or feared the sea as a She, and the moderns, who analyze her as a mere It. We stand with the poets.

To the poets, the sea is a little less than a goddess and a lot more than a mere chemical. She is neither Heaven nor mere earth, but the mirror of Heaven. The sea below reflects the sea above, and the unattainable stars. The sky is "the sea above," and the sea is "the sky below." The tangible, finite firmament below reflects the intangible, infinite firmament above. Thus Byron called the sea

Thou glorious mirror, where the
   Almighty's form
Glasses itself in tempests; in all time,
Calm or convulsed – in breeze, or gale, or
   storm,
Icing the pole or in the torrid clime
Dark-heaving, boundless, endless, and
   sublime –
The image of Eternity – the throne
Of the Invisible; even from out thy slime

The monsters of the deep are made; each zone
Obeys thee; thou goest forth dread, fathomless, alone.

To the mystics, she is something even higher than the mirror of the heavens; she is the mirror of the soul. The outermost sea mirrors the innermost fire.

But to the modern mind, she is only an it, she is nothing but trillions of tons of cold salt water. And water, in turn, is nothing but molecules of H_2O. According to this reductionism, this "nothing buttery," the mythmaker, the poet, the mystic, and the sailor are themselves only molecules.

The modern mistake is ridiculously simple: a confusion between what something *is* and what it's *made of*.

What the sea is, is infinitely more than what she is made of. The whole is more than the sum of the parts. She is more than her matter. Her "form" is more than her geometrical shape. Her form is her essence, her identity.

The mythmaker, the poet, the mystic, the sailor, the saint, and the child know her best because they know *her*; the scientist knows only many things *about* her. They know her personality; he knows only her blood type. A child knows "Mommy" better than a sociologist does.

Though the sea is not a goddess, she is more like a goddess than a chemical. So the pagan mistake of seeing too much in her is not as bad as the modern mistake of seeing too little.

Moses used her as a symbol for all matter. That cosmic stuff that God's Spirit blew on and formed in the creation story in Genesis – why did Moses call that "water"? If you watch the stormy sea pacing in its cage like a tiger, you will know. As you watch, you enter a time machine: you are watching a picture of the drama of creation. When He designed earth's stormy oceans, God painted a picture of His Spirit breathing His timeless passion out onto the heavy seas of time, on the first day of creation; and He hung this picture in

His nursery (earth) for His children to see once they would be born.

But what is it a picture *of?*

We may find a clue if we look more carefully at one of the old myths before we dismyth it. It is the geographical myth of Oceanus, the Encircling Sea.

The ancient Greeks' map of Oceanus was not physically accurate, but it *was* spiritually accurate. To them, this Ocean was an endless stream that flowed forever around the outside of the world, ceaselessly turning upon itself like a wheel. It was the end of the earth and the beginning of heaven. Ocean was boundless. If you were to venture far out upon it, you would come not to land but to a dreadful and chaotic blend of sea and sky, a place where whirlpools and yawning abysses waited to draw you down into the dark underworld from which there was no return. The Encircling Ocean was the dread Sea of Darkness.

That was Ocean. The Mediterranean, on the other hand, was only Sea. The land was surrounded by Ocean, but the Sea was surrounded by land. It was called the "Mediterranean" because it was in "media-terra," in the middle of the earth.

Spiritually, the Greeks were right.

There are two philosophies: one says that the sea is surrounded by the land, the other says that the land is surrounded by the sea. The sea is a natural symbol of spirit, mystery, and moreness, of what is more than us, of "transcendence." The land is a natural symbol of matter, reason, and "lessness," of what is less than us, of what we can control. So the two philosophies are: (1) that the land is surrounded by the sea, i.e. that matter is surrounded by spirit; or (2) that the sea is surrounded by the land; i.e. that spirit is surrounded by matter.

All premodern cultures believed the first philosophy. It seemed obvious. It was an innate intuition, either an innate illusion or an innate wisdom. According to this old philosophy, all that we see is only a little island, and is surrounded by an immense ocean of what we do not see. As Hamlet said, **"There are more things in heaven and earth . . . than are dreamt of in your philosophy."** The visible world is only the epidermis of reality, the tip of the iceberg. What lies beneath? Something more, something invisible, something mysterious, like the sea. And that sea occasionally breaks in and floods the land with storms: miracles.

Our civilization is the only one in history arrogant enough to claim to surround the sea, to explain the mystery, to reduce the unknown to the known, the uncontrolled to the controlled, the supernatural to the natural, spirit to matter, man to ape, soul to brain, love to lust, religion to psychology, God-who-created-man-in-His-own-image to man-who-created-God-in-his-own-image.

But the poets know better:

Roll on, thou dark and deep blue Ocean, roll;
Ten thousand fleets sweep over thee in vain.
Man marks the earth with ruin – his control

Stops with the shore; – upon the watery
  plain
The wrecks are all thy deed, nor doth
  remain
A shadow of man's ravage, save his own,
When for a moment, like a drop of rain,
He sinks into thy depths with bubbling
  groan,
Without a grave, unknell'd, uncoffin'd,
  and unknown.

Thus far our daily devotional reading from the pious atheist poet Byron. Or, to put the same thing in religious language, **"Arise, O Lord, let not man prevail."** (Psalm 9:19)

If the poets are right, if all other cultures except our own are right, then Mind and Spirit are not some little local latecomer, some evolutionary freak on some minor planet. Instead, the whole universe of planets and suns and galaxies, all matter, time, and space, is a little local thing, like an island, surrounded by an infinite sea: the Mind of God, our ultimate return address as Emily understood, in Thornton Wilder's "Our Town":

Jane Crofut
The Crofut Farm
Grover's Corners
Sutton County
New Hampshire
the United States of America
Continent of North America
Western Hemisphere
the Earth
the Solar System
the Universe
the Mind of God

It is at the beach, where sea and land touch, that the question arises: which surrounds which? Which philosophy is right? Is mystery only small and temporary? Are mysteries like love and beauty and death only things *in* our lives, like lakes? Or are we and our lives surrounded by them, as continents are surrounded by the sea? For instance, is love in us or are we in love? Are we bigger than love or is love "bigger than both of us"?

The answer can be found by asking another question: What do we get bored with? Whatever we get bored with is smaller than us. Our souls eat it and are still hungry, still empty. Our souls are bigger than what they eat. But there are some things we don't get bored with because we can't eat them; they eat us. They are too big to enter into us; we enter into them. Such things as joy, and love, and the beauty of great music, and the stars, and the sea.

If they are not bigger than us, they must be full of something like angels, messengers of Something or Someone who *is* bigger.

I think it very likely that the ocean is full of angels.

The Sea as a Koan

The sea keeps saying something. Her waves are her lips. What is she saying? What has she been repeating for millions of years? How can we learn her language? It must be too simple for us to understand, yet it is so wonderful that we never get bored hearing it.

I cannot die until I learn this language and read this riddle.

Haiku

The sea is a wordless koan.
It keeps asking
But never answers

The sea is my guru. I must solve its koan. That would be wisdom. Carl Sandburg says,

"The sea is large. The sea must know more than any of us." He is right. So is Alfred Noyes:

Who the essential secret spells
In those gigantic syllables –
Flowing, ebbing, ebbing, flowing –
Gathers wisdom past all knowing.
Song of the sea, I hear, I hear
That deeper music of the sphere,
Catch the rhythm of sun and star
And know what light and darkness are . . .
I hear, I hear, and rest content,
Merged in the primal element,
The old element whence life arose,
The fount of youth, to which it goes.

I too can say, "I hear, I hear, and rest content." My heart rests content with the sea when I hear her, even though at the same time I feel a restless discontent in my heart. Perhaps my heart is content with its own discontent. But my heart has not shared its wisdom with my mind; I do not understand the riddle.

Tolkien too thought our discontent had something to do with the angel music in the sea. In the creation poem in his *Silmarillion,*

The other Ainur [spirits, angels] looked upon this habitation set within the vast spheres of the World, which the elves called Arda, the Earth; and their hearts rejoiced in light, and their eyes beholding many colours were filled with gladness; but because of the roaring of the sea they felt a great unquiet. And they observed the winds and the air, and the matters of which Arda was made, of iron and stone and silver and gold and many substances; but of all these water they most greatly praised. And it is said by the Eldar that in water there lives yet the echo of the Music of the Ainur more than in any substance else that is in this Earth; and many of the Children of Iluvatar [the Creator] hearken still unsated to the

**voices of the Sea, and yet know not
for what they listen.**

For what *do* we listen? What secret do we hope to find? What does the sea signify? A thing as great as the sea must signify something. If the Mona Lisa is more than a million molecules of paint, the sea is more than a trillion tons of water. The sea is like the Mona Lisa in three ways: it is a great work of art; it is a face, a personality; and it is mysterious and ambivalent. What is the secret behind that smile?

Perhaps the secret is an angel, a spark of the divine fire shooting out from God's mind like divine electricity. Something both irresistibly beautiful and fearful, both wonderful and terrible, attracting and repelling at the same time, beautiful in its very fearfulness, fearful in its very beauty, terrible in its delight and delightful in its terror.

Its secret must be something huge, something ineluctable, unarguable, and irresistible – something like God. And therefore, like God, much too simple for us to understand.

ORENDA

The mystery of the sea is only the most obvious example of the mystery of all nature. For not only the sea but all nature contains a mystery. The mystery is this: Why does it fascinate us so? Why is it inexhaustible to our spirit? Why does nature make us happier than art? Why is there more in a stick and a stone than in a bat and a ball? Why is a forest of trees better than a forest of telephone poles or buildings?

We deliberately turn trees into telephone poles or buildings, yet we love the raw material more than the things we make out of it. Why? *Something* in us prefers cities, for we choose to build them and live in them; but something else in us prefers nature, and we all know that that "something else," whatever it is, is the happier of the two "somethings." What is that "something else"?

What happiness flows through waves and trees? Why does it make our blood race to its rhythms? Dylan Thomas says,

> **The force that through the green fuse**
> **drives the flower**
> **Drives my red blood.**

But what force is that?

An Iroquois would say it is *orenda*. A Taoist would say it is the Tao. But what is that? What is the spiritual power of matter?

Obviously, the reason why nature makes us happier than civilization does is because God made it.

> **Poems are made by fools like me**
> **But only God can make a tree.**

We could never have designed a thing as perfect as a wave. We can only design things like poems. But God writes poems with water, not words.

What power, what spiritual electricity, what strong magic, did the Creator put into the creation?

If we ever find the answer to that question, I think we will find it first in the sea. For there we find the most magic of all. All nature is radioactive with a mysterious spiritual power that we do not understand, but it shows itself most powerfully in the sea. We never feel more humble and helpless than when we face the sea – and we never feel happier.

The Joy of Ignorance

Maybe it's our very ignorance that makes us happy.

We can't predict Nature's lines. They are not geometrical. The exact shape of a wave is inexact. Where the lines of a sand dune, or of a rock, will appear is unpredictable. And that makes us happy. Why?

We know that these things have reasons, we just don't know the reasons. Only God does. So perhaps we love nature because we love to be reminded of our ignorance, and we love to be reminded of our ignorance because that assures us that there must be a God, since otherwise there would be no reason at all for most things! For *we* certainly don't know the reason for most things.

So nature does for us what Socrates did

for Athens: teaches us our ignorance, teaches us that we are neither God nor Nature but Humanity. We are neither Daddy nor Mommy but Baby. If Baby understood everything in the house as well as Daddy or Mommy did, Baby would not be Baby. And Baby would not be happy. Our ignorance in the face of nature makes us happy, deeply happy. We are happy when we know, as Baby knows, that Somebody Else is in control, somebody wiser and stronger. If Baby had Daddy's knowledge and power, Baby would not be happy.

Perhaps this is why technology has not made us happy (though it has wonderfully reduced one form of unhappiness: physical pain). In order to master nature, we have to exchange poetry for mathematics, we have to turn the light of reason into something narrow and powerful like laser light. And that narrows our happiness too. For when we tame nature, we have to ignore something in nature that we *can't* tame: the Tao, or *orenda,* or maybe even an angel. And it's *that* "something" that makes us deeply happy.

That's why we're more deeply happy in the presence of *untamed* nature, like the sea or the woods, than in the presence of tamed nature, like a salt water aquarium or a mowed lawn.

WATER

There may also be clues in what the sea is made of. God could have made the sea of anything, but He invented water.

Water is unique. If any one of its properties – its weight, its freezing point, its density – changed just a tiny bit, all life on earth would cease. Water is the primary requirement for life. Nothing else will do. Water is the planet's blood, and the sea is its heart, where that blood is received from the veins of rivers and given back through the arteries of evaporation.

In the Bible, water is used as a symbol of both life and death. God both heals and destroys by water.

God gave water the highest rank among

all material substances by creating it first. Only light came earlier.

It is a familiar idea in most religions that all creation reveals something in the Creator, as all art reveals something in the artist, and thus that all material things reveal spiritual realities, since their artist is Spirit. But the application of this principle to water is no longer a familiar idea, even to most religious people. We find it only in a few writers, like George Macdonald:

> I believe that every fact in Nature is a revelation of God, is there such as it is because God is such as he is. . . .
>
> The water itself, that dances and sings, and slakes the wonderful thirst – symbol and picture of that draught for which the woman of Samaria made her prayer to Jesus – this lovely thing itself, whose very wetness is a delight to every inch of the human body in its embrace – this live thing which, if I might, I

would have running through my room, yea, babbling along my table – this water is its own self its own truth, and is therein a truth of God. Let him who would know the love of the Maker become sorely athirst, and drink of the brook by the way – then lift up his head, not at that moment to the maker of oxygen and hydrogen, but to the inventor and mediator of thirst and water, that man might foresee a little of what his soul may find in God.

Seeing the Sea
with the Third Eye

A man like Macdonald *knows* water only because he *loves* water. Love gives him his third eye. We are no longer well versed in the art of third-eye seemanship; that is why the wisdom of a Macdonald sounds strange to us today.

This ancient art was practiced in all pre-modern cultures. It could be called the art of sign reading. It assumes that nature is not just a *thing* but also a *sign*, like a word, and therefore is not just to be looked-*at* but also looked-*along*. You look-along a sign, not just at it; you *read* the sign. But modern books about the sea always look *at* it instead of *along* it. So they miss its significance, its signing. They never learned its language, which is sign language. They're so busy imposing their own advanced scientific languages on nature that they don't listen to nature's own simple sign language speaking to them. When they see a tree waving in the wind, for instance, they never think the tree is waving at them. They think it's only being moved by air molecules. They think nature isn't *signing* but *spastic*. They are like unsocialized children who can't read body language.

This habit of sniffing at facts and missing signs pervades our whole modem culture. It is a doglike mind. Point to its food, and an untrained dog will sniff at your hand. For a few hundred years now we've been doggedly sniffing at the sea, experimenting with it, calculating solutions to scientific puzzles about it, conquering it intellectually and technologically, and figuring out how to reverse our pollution of it. And this is good. But it's *not* good that meanwhile we've been neglecting the ancient and honorable art of sign reading, because we've forgotten that we have another eye, an inner eye that is not just the eye of the face (the senses) or the eye of the brain (the mind) but the eye of the heart, the third eye.

The eye of the heart is the eye of love.

Even more than love, longing. The Germans called it *Sehnsucht* – a deep, mysterious longing for we-know-not-what, for some mysterious More. The heart sees farther than the mind can follow. **"The heart has its reasons, which the reason does not know."** (Pascal)

And this restless heart sees its natural icon in the restless sea. When we open our third eye, we immediately sense that the sea is a great sacred sign.

More than a sign, a meeting place. More like a door than like a window. A sign is like a window: light comes through windows, but not real things, unless the glass is shattered. But real things come through doors, and down highways, and up and down ladders. The sea is like a Jacob's Ladder linking earth to heavenly forces, a highway for angels ascending and descending. It brings a *smell* of Heaven to earth. A smell is more than a picture; it is a real, though tiny, part of the thing smelled. The air in the hallway actually brings molecules of fresh baking bread from the kitchen into the living room where we sit, puts them right into our bodies, into our noses and lungs.

And just as the bread we smell is an accurate foretaste of the bread we will eat, so salt air is in some way an accurate foretaste of Heaven. The sea brings some tiny molecules of Heaven to earth. Maybe Heaven is so big that even its smallest molecules are as big as the sea.

Or you could call it a kiss. The sea gives us a fleeting kiss, not just an abstract idea but a concrete foretaste, an appetizer, of the big kiss God will give us in Heaven. We have to become solid enough to endure that kiss; that's why life is so hard.

When the sea teases and ravishes us, it's foreplay. It's a lover's whisperings. And what does it whisper? "Taste and see the goodness of the Lord."

But you have to be very silent inside to hear that whisper.

The whole sea is a sermon and each wave

is a word. The whole sea is a symphony and each wave is a note. Its movement is ever-flow, never-stop, never-still, just like our own hearts. We can move with this sea, by swimming or by sailing or by surfing, but we cannot stop it, or stop in it. We can't build houses on it. It's like life itself: we can't capture it or stop it or hold it, any more than we can capture or stop or hold light in bottles. Not even word-bottles.

I think words are not *meant* to capture the answer to the sea's mystery, any more than they are meant to capture the strangely similar mystery of music. I think our question about the riddle of the sea is meant to be asked, but not meant to be answered. It is supposed to go on and on like the waves.

God, our hearts, and the sea, are all like that. Three inexhaustibles. No surfer in history has ever been heard to say: "Now I've had enough of waves." No lover will ever say, "Now I've had enough of her." And no saint will ever say, "Now I've had enough of God."

The Sea as Miracle

A miracle is an act of God. The sea is a miracle not only because it is an act *of God*, but also because it is *an act* of God. It's not just a thing, it's a deed, or a word. It doesn't just exist, it speaks.

Its voice is like the voice of God. As Elihu said to Job, **"God is speaking all the time, in one way or another, but we don't hear him."** (Job 33:14)

The voice of the sea must be like the voice of God because the voice of God is like the voice of the sea. For when the prophet Ezekiel heard God's voice, that is what he said: **"And behold, the glory of the God of Israel came from the way of the east; and his voice was like a noise of many waters; and the earth shined with his glory."** (Ezekiel 43:2)

The floods have lifted up, O God,
The floods have lifted up their voice,
The floods lift up their roaring.
Mightier than the thunders of many
   waters,
Mightier than the waves of the sea,
The Lord on high is mighty! (Psalm
   93:3–4)

Here is a test of your third eye. Why does the sea shimmer and shake?

The answer in the Bible (whose writers had wide-open third eyes) is this: For the same reason the prophets shook: they saw God's face:

The waters saw thee, O God, the waters
   saw thee;
They trembled, and the depths quaked.
   (Psalm 76:17)

Third-eye people know that the wind is God's breath because they listen to the sea teach them that. Like William Cullen Bryant in "A Hymn of the Sea":

The sea is mighty, but a Mightier sways
His restless billows. Thou, whose hands
   have scooped
His boundless gulfs and built his shrine,
Thy breath,
That moved in the beginning o'er his face,
Moves o'er it evermore. The obedient
   waves
To its strong motion roll and rise and fall.

You may wonder why I see the sea as a miracle, as something supernatural; isn't the sea the most natural thing in the world?

Of course the sea is natural, if anything is. The sea is the most "worldly" thing in the world: it is four-fifths of the world. Yet "the sea is another world," as Anne Stevenson wrote. I think she meant by that not that the sea is extraterrestrial, like an alien or an angel, but that it is superterrestrial, totally terrestrial, terrifically terrestrial. And that's why it's so supernatural. It's so supernatural because it's so natural.

For all of nature is a miracle, in fact the primary miracle. And the sea is the closest part of nature to the whole of nature. So the sea is the most miraculous thing we see. People come to disbelieve in miracles only when they forget the most massive and obvious of all miracles, nature itself; and that's hard to do when you live by the sea.

The sea has a personality because of her border with the land, which defines her. Nature has a personality because her border with the supernatural defines her. If nature is only "all there is," then how can she have such a distinctive personality? Why is she so tangy and salty and unpredictable and lively? Why is she a *she?* If there were no *He,* the word "she" would be meaningless. Nature must have an Other, a Lover.

Here is a simple proof that nature is a miracle. A thing that can only be caused by God and not by nature is a miracle. But nature is a thing that can only be caused by God and not by nature. Therefore nature is a miracle. Nature can't create herself. (Even God can't create Himself. Nothing can. God is not created at all.)

Perhaps some events that we call miracles

really come from some part of nature that we just don't know about yet, rather than directly from God. But nature as a whole could not possibly have come from any of her parts. Therefore nature as a whole is a miracle.

If you want to understand this, live by the sea and you will see God. For God surfs!

**Thy way was through the sea,
Thy path through the great waters;
Yet thy footprints were unseen.** (Psalm 77:19)

The sea is the earth's most remarkable icon for God:

Like God, she is "beauty ancient yet ever new," unchanging yet ever changing, eternal yet dynamic, immovable yet moving, timeless yet alive, formless yet full of personality, infinite yet definite, fearsome and wonderful, full of life and Mother of all life, source and master of life and death, maker and breaker of living things.

And like God, she is wild. (God is a wild man; if that makes no sense to you, you have never met Him.) Thoreau says, in *Cape Cod,* **"The ocean is a wilderness reaching around the globe, wilder than a Bengal jungle and fuller of wonders."**

The sea is also Godlike by being the absolute. Land is relative to sea, not vice versa.

The sea is also Godlike by being vast and deep.

And by being self-effacing. It lets you just sit there and watch it, or ignore it. It only occasionally shouts "I'm here!" in storms. It's the taken-for-granted background to everything in the world, and it lets itself be forgotten for a long time. But not forever.

The sea is also like God in being a perpetual presence. It is always there, the omnipresent background to our lives. Whatever happens to us, happens to us on the water planet. Every event in human history has been surrounded by the sea, as every event in our lives is surrounded by God. However trivial or ugly the *plot* of a life may be, the *setting* is always deep and beautiful.

Another reason why the sea naturally

symbolizes God is because its matter naturally symbolizes spirit. It seems to be composed of matter that is only a split second away from becoming spirit, or spirit that just solidified into matter a split second ago. Loren Eiseley, in *The Immense Journey,* says, "If there is any magic on this planet, it is contained in water . . . remarkably like the mind."

If water is like mind, then mind is like water. And Eiseley sees that half of the equation too:

> As for men, those myriad little detached ponds with their own swarming corpuscular life, what were they but a way that water has of going about beyond the reach of rivers? I too was a microcosm of pouring rivulets. . . . I was three fourths water, rising and subsiding . . . in my veins. Thoreau, peering at the emerald pickerel in Walden Pond, called them "animalized water."

As a chicken is an egg's clever way of making more eggs, a man is water's clever way of invading the land.

Why is the sea is Godlike in so many ways? Because it emerged straight from God's mind, and His mind is still upon it. It is a mirror. Of course Mind is spiritual while both the sea and mirrors are physical, yet both the sea and mirrors are natural symbols of mind.

The mind is most visible in the eyes, "the windows of the soul." If you want to see a person's soul, look into his eyes. Look *into* them, not *at* them; look-along his outer eyes with your own inner eye. If you do that, you will see his mind. Now do the same thing with the sea: look into her eye, and you will see the Mind of God.

Of course you can't see God, or any mind, with your outer eyes. And if you could, you would be blinded. Even the sun blinds you if you look at it directly. But you can observe it, during an eclipse, by looking at its image reflected on paper. You can do a similar thing with God. The paper here is the sea.

When you sit by the sea and watch the waves with both attentiveness and peace, both alert and calm, without either laziness or agitation, then your inner eye will open and you will see the mind of God reflected in His great earthly mirror. For He designed it precisely for that purpose: to reflect Him to you.

Moving water is holy water. It exorcises the evil spirit of atheism. It baptizes the unconscious. It is more powerful than any argument. That's why there are few atheists who live by the sea, or even by a swiftly moving river or brook.

Take a baby from birth. Put him near the water. Let him grow up on the shore. Let the water be his only teacher. Let the waves be their words. I will wager that he will not grow up to be an atheist.

What could an atheist *do* with the sea? How could he *place* it? For him, no Mind designed it, no Artist loved it into existence. It is not art, it just is. It does not mean, only be. It has no place, no address, no home in the Mind of God. No Heavenly Father gave it to him as a gift. What a terrible moment that must be for an atheist – when he feels great gratitude for the gift of the sea and there is no one to thank for it.

For most of us, the most terrible moments in life are the moments of grief; when hope is tested. For an atheist, I think the most terrible moments must be the moments of joy, when there arises from his heart the wisest and best feeling anyone can have – cosmic gratitude, praise for his very existence and that of the whole universe and he has to believe that that feeling is false, stupid, wrong, pointless, and out of tune with reality, since reality is nothing but chance and gravity and molecules. That feeling of cosmic gratitude and praise was a *temptation* to be "religious." Imagine how horrible it must be to feel that gratitude is a temptation! That's almost like believing that goodness is evil. It's like waking up one morning to see that the sun is a Black Hole.

Wonder and gratitude can be triggered by anything – a bird or a baby or a buttercup; but there is nothing more potent at raising swells of praise in our innermost sea than the waves of the outermost sea. **"Deep calleth unto deep at the noise of Thy waterspouts; all Thy waves and Thy billows have gone over me."** (Psalms 42:7) Advice for atheists: don't live near the sea, for if you do, you will have to suppress your own heart. Don't be too friendly with the sea, for if you do, you won't be able to be friendly with your own heart; you'll have to scold it for being a fool.

When you live by the sea, everything changes, and the change is the same as when you believe in God: you are never alone. There is a Greater Presence next to you every minute. You have to take account of this Presence every day, at least unconsciously. When a landlubber asks what the weather is like today, he only means to ask about quantities of heat, precipitation, or wind velocity. But when a seasider asks that question, he means: What is She doing today? What's Her mood? What's Her will? How is She acting? You always have this large, unpredictable wild animal in your neighborhood. It's like having a 500-pound mother-in-law living in your back yard.

MUSIC

Once, when I was very young – too young to remember how old I was – I think I heard what the sea was saying. I think I heard the secret of the sea. I read the riddle. I cracked the code. I understood the secret language. The language was music.

It was angel music, I think. I know it was not human music. It was a music that was there under all human music, like an organ point or pedal point under all melodies, like the bagpiper's drone beneath the piped tunes. Only the child in us can hear that sound. It is too deep to register on the adult ear, which scans only for the shriller levels of sound. When I was very little, I understood that the whole planet was a gigantic musical instrument, perfectly tuned, and the sea was its first bass. (That's how I word it now; I had no word, or need for words, then.)

Who played on this instrument? Angels. It must be angels.

Children see them more readily than adults, and women more often than men, I think, because their receiving sets are more tuned to the angels' wavelengths. The angel station is not Analysis but Intuition.

The music is playing all the time. All the intermittent musics on earth are surrounded by the perpetual music of the sea, and the music of the sea is surrounded by "the music of the spheres."

Why doesn't God let everyone hear this angelic music all the time, if it is so beautiful? Maybe He does, but we don't hear it because we let worldly wax grow in our ears. Or maybe He doesn't; maybe it's not our folly but His mercy that insulates us from it; maybe God puts cotton in our ears because beauty so big would drive us mad (the sea almost does, already); because if we heard those cosmic chords, so tremendous and remote, perpetually shaking the skies and making Him that sitteth in the heavens to laugh, we would be unable to eat, sleep, reproduce, or survive; because it would be an unbearable weight of glory that would crush our spirit like a locomotive crushing a bug, a fire that would bum up our mind as a furnace burns a moth. Maybe we would just disappear! Maybe the only way to survive in this angel-haunted universe is to be deaf and dumb.

Surely it would squash us if we shared the angels' minds, if we saw all they saw. Just imagine one thing: imagine you could see all the differences every one or your prayers made, all the echoes of the tiniest notes of your prayer music in everyone else's life down through all future generations. Those echoes do not peter out but increase and reverberate as they move into the future and mingle with other echoes. If we saw that, I think we would then know the literal truth of Dostoyevski's saying that **"we are each responsible for all,"** and we would be so awed with our power and responsibility that we would never be able to get up off our knees again for the rest of our lives.

Even a faint echo of one note in this cosmic music, the sea's note, heard by the paltry poet, the moldering mystic, or the semi-saint in our tiny hearts, is too much for us to handle, to understand, or to tell. It's like having a secret lover that you can't reveal and can't conceal. A very few of us can make this music into Brandenburg Concertos or Choral Symphonies. But none of us can lay hands on it any more than we can lay hands on a hurricane. It lays its hands on us. It plays on *our* organs.

And I think I heard this hurricane once, when I was very young. Was it in a dream or was I awake? I don't know. The difference between dreaming and waking is less clear than the difference between that music and human music. I don't know how I heard it, and I don't know how I remember it. But I know that when you hear the music of the angels you can never believe it is only human. And you can never forget it.

Probably Adam and Eve heard it all the time in Eden. Probably we vibrate to it because in something like our "collective unconscious" we dimly remember the echoes of Eden's love song reverberating down the corridors of the centuries. It's the music of Home heard by the homeless.

But how is it different from ordinary music? If it's so radically different, how can I call it 'music' at all? And why do I insist it was angel music if I didn't see an angel?"

The answer to both questions is in the music itself. All other music makes me thrill with emotion only if it thrills with emotion, but this music makes me thrill with emotion because it *doesn't*.

Let me try to explain. Thoreau once described the sea (in *Cape Cod*) as **"naked Nature, inhumanly sincere."** That's a good description of an angel: naked spirit, inhumanly sincere. If there's anything an angel is, it's nakedly sincere: pure, simple, perfect, a million miles from the slightest suggestion of uncertainty, complexity, animal passions, or human compassions. The sea has no compassion – any sailor will tell you that – and I think angels too have no compassion, though they have dazzling understanding and pure, fiery goodwill. They are "inhumanly sincere." By "sincere" I mean that both the sea and the angel are simply what they are, nothing else.

I must confess I'm not sure I'm right about this because I'm not sure the sea is emotionless, and I'm not sure angels are emotionless either. For that matter, I'm not sure whether God is emotionless. (He's beyond *human* emotion, of course, but there may be other kinds.) All three are mysteries. But all three are similar. The irony is that in all three cases, what is beyond human emotion can inspire the greatest human emotion.

The composer of this music is God, of course; the angels are only the players, and the planet is their instrument. That's probably part of the answer to the riddle of why I can sit and listen to the music of the sea forever. Is there any man – made music you can listen to forever? But this music you *can* listen to forever. Therefore this music is not man-made music. That's not fantasy, that's good logic.

It's also good science. It's empirically testable. God's water music has detectable effects: it changes you, it heals you, it refreshes you, more than man-made music does, for the magic of man-made music pales after enough repetitions, even when it is long and complex; but God's water music never gets boring even though it is extremely simple.

This is a strong clue that the sea's secret ingredient is God's Spirit. For nothing finite is inexhaustible; everything finite eventually gets boring. Perhaps God's own Spirit is the spiritual salt of the sea. He likes to exalt His subordinates, so He probably using the little created spirits we call angels as His instruments. (I say "little" spirits. They are little compared with God, but each one is larger than the universe.) Just one angel would do, I think, one angel who fell into the sea, or fell in love with the sea.

WATER IN LOVE

Water and air are obviously lovers. The wind endlessly caresses the sea's hair, and in response the sea stirs and purrs, and smiles. Sometimes her response is a wild orgasm: a storm.

Many ancient cultures see Air as Breath, Breath as Life, and Life as Spirit; thus the word for "air" and "spirit" are often the same word: *ruah'* in Hebrew, *pneuma* in Greek.

It was Spirit that "blew" over the face of the "waters" at the very beginning of creation (Genesis 1:2). Their love formed the world, and continues to form it, as a man's and a woman's love forms a child, and continues to form it.

All the little miracles in the world are echoes of this primal love. They are storms on the sea of our lives made by the wind of God's Spirit breathing kisses.

Sea, what a peerless matchmaker you are! How many romances have you smoothed? How many souls have you saved from loneliness? How many couples have you seen wondering at each other because they have learned to wonder by wondering at you? How many have sent their hearts into each other because they sent both their hearts into you?

Matchmakers are always women. So are seas. In all languages, "sea" is feminine. She was the womb of all life on earth. She was the mother of the moon, her firstborn daughter tom from her Pacific womb when she was young and hot. These two, Mother Sea and Daughter Moon, still wave the baton and set the tempo for the life-music played by the body of every woman on earth. A woman's periods do not *match* those of the moon and the tides; they *are* those of the moon and the tides. Human life's maternal cycle is monthly

– *moonthly*. The tide charts of a woman's inner sea of blood are the same as the tide charts of the earth's moon-moved outer seas of water. Earth's seas overflow once a month too, when sea and moon are aligned.

The sea's periods are both regular and irregular, like a woman's. This is the pulse of life itself: ebb and flow, storm and calm, waxing and waning, twister-like turbulence and lake-like tranquility. Both She's are regularly irregular, predictably unpredictable.

All the poets know the sea is a woman. Swinburne, for instance:

> I will go back to the great sweet mother,
> Mother and lover of men, the sea . . .
> Born without sister, born without mother,
> Set free my soul as thy soul is free.

You can test a man's attitude toward women by asking him why the sea is a woman. Some will say because she is treacherous, others because she is beautiful, mysterious, and wise.

I pity those who live near the shore but in perpetually cold climates, like Oregon. To look longingly at the waves without ever being able to enter them, to mingle with them, to *become* the sea; to be forever on the outside of nature's most beautiful mystery – that must be maddeningly teasing, like never being allowed to make love to your spouse.

The sexual imagery folds back on itself like origami; for sometimes the Great Mother Sea becomes a man and the land becomes a woman. The phallic nature of those white waves that thrust their foaming tips into the receptive caves of shoreline rock is embarrassingly obvious. And unlike other men, the sea does not weary after only one orgasm, but persists, like the righteous in Heaven in an endless ecstasy of pure, eternal lovemaking.

And then the image folds back again, and the little white waves that I am watching as I am writing this become a woman's hair swirling around a man's neck, teasing the land as a woman teases a man.

Sea Tease

Endlessly the teasing sea
Swirls her hair around the rock,
Bares her heaving, briny breast,
Taunts and tilts her hair-spray head,
Parts her lips to kiss the sand,
Then dances, laughing, back again.

No more can she stop this tease
Than the rock can ever start
To leave his adamantine place
To join her dance, and sway, and weave.
Mother to all earthly life,
Sister to the spirit wind,
Daughter of earth's fiery core,
Mistress to the manly rock,
Mirror to the shining stars,
Draw my heart into thy waves again.

ALIVENESS

The sea can symbolize God because it can symbolize spirit, and it can symbolize spirit because it is alive.

It doesn't just *contain* life, like a glass aquarium, but it is itself alive, like a large animal, almost a person. It is more like a somebody than like a something.

It *breathes*. The surf is its inhaling and exhaling. So are the tides. A wave is a hand reaching out to grab you, or to caress you. Perhaps to tell you something.

The sea's ceaseless movement has often been likened to breathing. But maybe it's not the sea that's breathing but the land. Maybe the water is the breath, not the breather. After all, water is the life of all living things, and the life is in the breath. Maybe water is to land what air is to us.

More than just *alive,* the sea seems almost *conscious.* She seems to have the power to read our diverse needs and supply them, just like Mother. The same sea will inspire and excite us when we are dull and depressed, and then console and pacity us when we are nervous and worried. How does she do that? Mere matter cannot read minds; but the sea seems to read our mind; so the sea seems to be more than mere matter.

Waves are lips. Lips can kiss, or speak, or bite. Gentle little waves kiss the children who play in them. Larger waves kiss the surfers who ride them. Bigger waves than we can handle bite us. But what do they speak?

All waves speak, but they speak in tongues, and we can't interpret their speech. That's probably because it's too simple, like God's. Maybe all they're saying is I LOVE YOU, I LOVE YOU, I LOVE YOU, I LOVE YOU, I LOVE YOU until the end of time. Like God.

The sea is never dead. Even on a calm day, when waves are not breaking, the sea is alive because she is never still. There is always that twinkle of the light in the air reflected on the face of the water, like the twinkle in the eye of Santa Claus. Even on the calmest days she is a quivering jelly, like Santa's belly. When there are no waves, there is still a swell, and a sloped shoreline of rock will create eddies of agiation.

The sea is constantly waving at us. Waves are verbs, not nouns. The sea does not "have waves," the sea *waves*. A wave isn't just something that happens by itself as the wave of a human hand doesn't happen by itself. It is a sign. One who waves has something in mind. What does the sea have in mind when she waves at us?

THE DARK SIDE

Like us and unlike God, the sea has a dark side. "To err is human; to forgive, divine," but the sea never forgives. It is the world's largest graveyard.

Water naturally symbolizes death. Sometimes this is merely terrifying, but sometimes it is also exultant, as in Emily Dickinson:

Exaltation is the going
Of an inland soul to sea,
Past the houses – past the headlands –
Into deep Eternity.

Water can also symbolize chaos. "The waters" were what God's Spirit tamed and formed at creation's first wrestling match. God breathed His Spirit out over the waters, the primordial chaos, and turned it into ordered cosmos. Many cultures have a "universal flood" story, like Noah's, because they are haunted by the nightmare of the formed cosmos sinking back into this formless watery chaos.

This chaos can be mental chaos too. The mind as well as the body can drown in the sea. If you have the habit of staring into it like a lover into the eyes of the beloved, its eye can hold you like Medusa. The spirit of the sea is far stronger than the human spirit, and captures it easily, especially in storms, the most exciting of all the sea's charms and also the most destructive.

Why do we find the most destructive things the most captivating and enrapturing things?

Not all of us do. Poets do, practical people do not. Sailors do not love storms. Those of us who only make pilgrimages to the sea love her best when she is stormy, but those who live on or near her hate and fear her fury like the fury of a tyrant. But she gives the poet, and the poet in all of us, a strange, deep pleasure that is a kind of pleasant terror; not just a contentment and satisfaction but a wonder and fear that we find more delightful than the contentment that calms that fear. For that fear is not a fear for our personal safety but simply a fear at her size and majesty. Longinus called this fear "the sublime."

We feel this wonderful fear most when we are alone. The sea looks tame when seen from a crowded beach full of blankets and umbrellas and chairs; but the same sea looks very different at night when the beach is deserted and you are alone. The water seems to leap up and bow down. It rises and falls like a drunken sailor. It is unpredictable. Little waves seem big at night when you are alone. And this is when we love the sea in a peculiar way, when we fear it most.

Why do we love what we fear?

Perhaps we love the tumultuous waves, we storm-lovers, because we love mirrors. For we ourselves are tumultuous waves. Our hearts love crashing surf because the lovers' hearts are full of crashing surf. The sea is our "know thyself' mirror, our heart's biographer. And on that inner sea of the human heart there has been an inner storm ever since we left Eden.

We moderns feel a much greater need for this sublime terror because our safe, technologically advanced world no longer offers us many fears and terrors. We fear boredom more than we fear fear. (I think that's a good part of the reason we fight wars: out of boredom.) We are not made for boredom, we are made for storms. Our inner tumult seeks its outer kin, the sea is the soul's long-lost sibling.

And when we find it, we know that we *fit*, that we are part of the same cosmic order as the sea, that we are not ghosts or aliens, that we belong here. We and the sea are in the same play, we as the characters and the sea as the setting.

Carl Sandburg must have sensed something like that when he wrote, in "The Young Sea,"

The sea is never still.
It pounds on the shore
Restless as a young heart,
Hunting.
The sea speaks
And only the stormy hearts
Know what it says.

If there is any heart that can understand the sea, it must be a young heart, a restless heart, a hunter's heart. "The heart is a lonely hunter."

The sea is not evil, but it is fearsome and dangerous. So are we. Its "dark side" fascinates us not because we long to see death and destruction, or because we long to die or destroy. Just the opposite: it is *good* people who love the sea when she is angry.

Why? Good people do not love human anger; why do they love the sea's anger? They are not fascinated with murder and mayhem and mass destruction when perpetrated by man; why are they fascinated with these things when perpetrated by a hurricane?

I think it is not our pride and arrogance that loves destructive storms, but exactly the opposite, our humility and piety. We love the great and terrible sea because we know we need humility, and she supplies that need. Man has become so big, and nature so small, so emptied of mystery by science, of danger by technology, and of poetry and myth by rationalism, that we no longer feel ourselves confronted by something we are even *tempted* to worship as a god or goddess, something greater than ourselves. What a loss! I say if you are not even *tempted* to worship the sun, you do not truly know it. It is to you only a large, efficient radiator.

I don't mean to be impious. I believe the Commandments, especially the one forbidding idolatry. But what would you say if you met someone who didn't need the Commandments, not because he was holy but because he wasn't tempted – for instance a healthy young man who was never even tempted to lust when he saw an incredibly feminine woman? I'd pity him. Where is the glory of conquering temptation if there is no temptation?

We know much more about the sea than our ancestors did, and that puts us in danger of forgetting Lesson One, how little we know. The sea is one of the few remaining reminders of this, and that is why she gives us the pleasure of humility. (Yes, I said "pleasure.") We were designed for that pleasure, for the big pleasure of being a small child in a big house.

When we are at the shore, we remember who we are and what we are doing: we are little children playing with sand castles. Sir Isaac Newton, one of the most spectacularly intelligent human beings who ever lived, said shortly before his death, **"To myself I seem to have been only like a boy playing on the sea shore . . . while the great ocean of truth lay all undiscovered before me."**

The Sea as Toy

I know no better image than that one to summarize our place in the cosmos: little children at the beach playing in the waves. Waves of life, of energy, of the Tao; waves of evolution, of predestination, of divine providence; waves of beauty, truth, and goodness, waves of the stuff God is, waves of God-stuff.

To Baby, the sea is the great playmate, the great Mommy. At the beach, Mommy and Baby smile deep, understanding smiles at each other. Little children know what the beach is for: it's a big sandbox, a toybox. Look at what children always do in the surf: they go wild-happy playing in the skirts of the Great Big Mommy. They are wiser than adults are, for they know what the universe is for: it's for delight. That's why any great artist creates. What do you think God is, an engineer? A lawyer? He's an artist!

And a toymaker. There is no need to forget what we knew as kids – that the world is God's toy for us to play with – when we become adults and learn that it is also a vale of tears, a vale of soul-making. There is no need to lose this sense of being small and happy, and becoming large and worried instead. Both toys and tears shape souls.

The sea is the perfect toy. It's unbreakable and unloseable, always available and always alive. It plays and plays with you without ever getting tired or bored. It dances with you and wrestles with you and boxes with you and tosses you around. It's just dangerous enough to be exciting. And you never have to put it away when you finish playing with it.

The surf can make us all children again in five seconds if we only let it. Think a truly radical thought: think what a revolution it would be if everyone on earth played in the surf once a week. How much depression and suicide, how much hatred and violence, how much resentment and anger and envy and boredom and addition, how many wars and murders and plots and tyrannies would just go out like a candle in the water? The sea is a peacemaker. How can surfers be warmongers? How could anyone drenched with the wisdom of playwater ever come up with this brilliant idea, the idea that has moved so much of our history? – "Hey, it seems we've got problems. Let's deal with them this way: let's dress up in funny uniforms and go out and kill each other."

The Fountain of Youth is not a myth. It exists. Ponce de Leon looked for it in Florida. That was not a mistake: you can find it in Florida because Florida has beaches. His mistake was looking for it inland.

The sea really is a fountain of youth. When I run into her great mammalian arms, I become Baby again. Mommy's arms cradle me, rock me, and wash me. The dirt of hurt and dreariness is washed away and I find underneath a surprisingly babylike skin. Baby keeps soiling herself, but Mommy changes her diapers: the tides keep cleaning out our pollution. I don't remember how it felt to have my dirty diaper changed – until I run into the sea. Then I know. Suddenly, I am *cleaned* Baby, purring and cooing and chortling in Mommy's strong arms. Nothing else can be as Mommy as the sea. Nothing else can make us feel so Baby as the sea.

Baby is the sand, and Mommy plays peek-a-boo with Baby. The sandpipers follow this peek-a-boo back and forth. Mommy's hand reaches out in waves of tickling. Baby giggles at her touch. If the sand harbors a hidden consciousness, it must feel what Baby feels: it must be giggling under each teasing wave. Baby is tickled pink that the great salty hand that hides such awesome power would temper its thunder to a tickle. How like God she is!

When you get cut by the rocks of life, you can run to Mommy's soft refuge. Waves have no corners. Mommy will kiss all your psychic boo-boos with her mighty magic mouth. At night, she becomes the blanket that wraps herself around Baby earth as Baby turns in his sleep. Baby has come a long way, for Baby has been rocketing through outer space all day. He is a starship, and his name is the Enterprise.

The Sea as Healer

The sea's power of healing is known to every age. This doctor is always in her office, always sees patients, and they always emerge healed. How does this work? What is the saliva in her kisses that heals the sores of our souls? Why does salt water make us happy? We still have not read our riddle.

Let's go back to our primary data, our simplest beginning: personal experience. What happens to me at the sea?

In a word, God happens. I hear God. The sea is His tongue. He speaks to me there. He assures me that He is still in charge, that there is beauty forever beyond the reach of human folly and selfishness. We can chip away all the rocks but not the sea.

Whatever happens to humanity, to society, to history, the sea remains. Civilization fell into a Dark Age, and may fall again; but the sea remains. Her waves do not descend into a Dark Age. Society falls into decadence, but the sea always remains innocent and uncorrupted. Her waves do not fall into decadence. They remain pure, a salve for society, an oasis from history, always available in any century – like God. If the world becomes Brave New World, the sea will not follow it.

As we slouch ever closer to Brave New World, the sea becomes more important than ever before. She is the great antidote to the poison Solzhenitsyn called "hastiness and superficiality." Even if modem life is trivial and shallow, she is serious and deep. We create an unholy racket, she creates holy silence.

Her silence is not just a nice thing for poets and thinkers; it is a necessary thing for sanity and wisdom. For as Kierkegaard says, **"If I could prescribe only one remedy for all the ills** of our modern world, I would prescribe silence. For even if the Word of God were spoken in our world, it could not be heard, for there is too much noise. Therefore, *create silence.*"

The sea creates silence. In five steps:
(1) The sea teaches me to listen to it.
(2) By listening to it, I learn the art of listening.
(3) By learning to listen, I learn to listen to God.
(4) By listening to God, I learn to listen to others, to the sea-depths in others, where the Spirit of God moves like swells.
(5) And only when I listen to others do I understand myself.

When I am with the sea, I am with myself.

We cannot hear God's voice until we become silent. For that voice, though larger than the universe, is also softer than a whisper. It is "a still, small voice." And that is more fearful than a hurricane:

And there he [Elijah] came to a cave, and lodged there; and behold, the word of the LORD came to him . . . "Go forth, and stand upon the mount before the LORD." And behold, the LORD passed by. And a great and strong wind rent the mountains, and broke in pieces the rocks before the LORD; but the LORD was not in the wind. And after the wind an earthquake; but the LORD was not in the earthquake. And after the earthquake a fire; but the LORD was not in the fire. And after the fire a still, small voice. And when Elijah heard it, he wrapped his face in his mantle. (I Kings 19:9–13)

You cannot argue with this voice, any more than you can argue with the sea. This voice is Truth.

We have filled our lives with noise, though this has not made us happy. Why have we done this? Precisely to avoid hearing this voice.

The Celtic monks of the Middle Ages made the opposite choice. They wanted to hear this voice. That is why they fled the sweet, green Irish countryside for the wildest, fiercest places they could find: barren, wind-swept, wave-lashed rocks in the sea: because they heard God there, and felt the vertical life surging through them, the pillar of fire from Heaven.

There is much more reason for the flight to the sea today, for our souls are getting older and shabbier and deader: passionless and heroless and saintless and poetless and petty and flabby and yuppiefied. How can such a soul see itself as the point of the whole great drama of evolution? From ameba to slug to snake to ape to man – and back again to the soul of an ape, snake, slug, or ameba in a man's body.

That's the deepest reason for alcohol addiction, drug addiction, violence addiction, and sex addiction. Deep down, we know our souls need something wild, something dangerous, something that makes us feel alive.

The sea is wild and dangerous and makes us feel alive. It's the last untamed place on earth.

TIME AND POWER

Technology is the most obvious and per-
vasive new feature of our lives, the one thing
we are very, very good at, the defining feature of our modem civilization. Machines are
good things, of course, and the artificial is
natural to us; after all, our own hands are
tools. But they are less deep and less real and
less alive than we are, and we are becoming
more like them, less deep and less real and
less alive. We have made our good servants
into our bad masters.

We did this by giving our machines two
precious things, time and power, the two
things we designed them to give *us* more of.
But everyone knows it has worked the other
way: we have less and less time the more
time-saving devices we have, and we feel

more impotent, more harried and hassled than our pre-technological ancestors. Our time-saving devices have turned us into worried slaves instead of leisured masters.

The sea is a powerful antidote to this. For the sea gives us time, when we give time to it. Like Christ, she multiplies the loaves and fishes of our time if we will only give them up first. The same is true of power as is true of time: she gives us more power and control over our lives if we first give it up, if we give up trying to control our lives, if we just sit there quietly for an hour a week and watch waves. When we do that, we emerge stronger, because we have left our obsession with strength back on land.

By a wonderful paradox, the same sea that restores our inner passion also restores our inner peace. Like God. For this wild thing is also a supremely peaceful thing. Like God. And she gives us a peace the land cannot give, a passionate peace. Like God.

That's why it's a mystery.

ABOUT THE AUTHOR

Peter Kreeft, Professor of Philosophy at Boston College, is also the author of over forty books, including *Socratic Logic* from St. Augustine's Press, as well as *The Philosophy of Tolkien*, *You Can Understand the Bible*, *A Summa of the Summa*, *How to Win the Culture War*, *Handbook of Christian Apologetics*, *C.S. Lewis for the Third Millennium*, *Three Approaches to Abortion*, *Love Is Stronger than Death*, *Socrates Meets Jesus*, *Ecumenical Jihad*, and many others.

If you talk to him, however, Peter Kreeft will get the biggest gleam in his eye not about his books but about surfing.

SABON

This book was set in Sabon. A descendant of the types of Claude Garamond, Sabon was designed by Jan Tschichold in 1964 and jointly released by Stempel, Linotype, and Monotype foundries. The roman design is based on a Garamond specimen printed by Konrad F. Berner, who was married to the widow of another printer, Jacques Sabon. The italic design is based on types by Robert Granjon, a contemporary of Garamond's.